Otfried Hankel

Early Jurassic palynomorphs from the Adigrat Sandstone in well Abred-1, Ogaden, Ethiopia

GRIN Verlag

Bibliografische Information der Deutschen Nationalbibliothek:

Die Deutsche Bibliothek verzeichnet diese Publikation in der Deutschen National-
bibliografie; detaillierte bibliografische Daten sind im Internet über http://dnb.d-
nb.de/ abrufbar.

Dieses Werk sowie alle darin enthaltenen einzelnen Beiträge und Abbildungen
sind urheberrechtlich geschützt. Jede Verwertung, die nicht ausdrücklich vom
Urheberrechtsschutz zugelassen ist, bedarf der vorherigen Zustimmung des Verla-
ges. Das gilt insbesondere für Vervielfältigungen, Bearbeitungen, Übersetzungen,
Mikroverfilmungen, Auswertungen durch Datenbanken und für die Einspeicherung
und Verarbeitung in elektronische Systeme. Alle Rechte, auch die des auszugsweisen
Nachdrucks, der fotomechanischen Wiedergabe (einschließlich Mikrokopie) sowie
der Auswertung durch Datenbanken oder ähnliche Einrichtungen, vorbehalten.

Imprint:

Copyright © 2013 GRIN Verlag GmbH
Druck und Bindung: Books on Demand GmbH, Norderstedt Germany
ISBN: 978-3-656-35740-7

This book at GRIN:

http://www.grin.com/en/e-book/208243/early-jurassic-palynomorphs-from-the-
adigrat-sandstone-in-well-abred-1

GRIN - Your knowledge has value

Der GRIN Verlag publiziert seit 1998 wissenschaftliche Arbeiten von Studenten, Hochschullehrern und anderen Akademikern als eBook und gedrucktes Buch. Die Verlagswebsite www.grin.com ist die ideale Plattform zur Veröffentlichung von Hausarbeiten, Abschlussarbeiten, wissenschaftlichen Aufsätzen, Dissertationen und Fachbüchern.

Visit us on the internet:

http://www.grin.com/

http://www.facebook.com/grincom

http://www.twitter.com/grin_com

Early Jurassic palynomorphs from the Adigrat Sandstone in well Abred-1, Ogaden, Ethiopia

Otfried Hankel

ABSTRACT

Palynomorphs have been recovered from the Adigrat Sandstone in the subsurface of the southeastern Ogaden region of Ethiopia. Core samples from well Abred-1 have yielded rich and well-preserved assemblages which are characterized by the dominance of the pollen genus *Classopollis* (average frequency 93%). Significant accessory elements are *Anapiculatisporites dawsonensis, Ceratosporites helidonensis, Foraminisporis tribulosus, Nevesisporites vallatus, Retitriletes semimuris, Araucariacites australis, Callialasporites turbatus,* and *Perinopollenites elatoides*. The microflora can be correlated with the lower part of the *Corollina torosa* Zone of Australia. It is concluded, therefore, that the studied section is of Hettangian or Sinemurian age. Continental deposition of the Adigrat Sandstone occurred probably under hot and dry climatic conditions. The presence of small numbers of acritarchs indicates a restricted marine influence.

Introduction

Oil exploration wells drilled in the central part of the Ogaden Basin of Ethiopia have penetrated an up to 1600 m thick continental sequence below the marine Jurassic-Cretaceous succession. The continental strata are of Karoo age and have been subdivided into four units. In ascending order these are the Calub Sandstone, the Bokh Shale, the Gumburo Sandstone, and the Adigrat Sandstone (WORKU, 1988).

The term Bokh Shale is here replaced by the term Tulli Shale because the type section of the Bokh Shale in well Bokh-1 is of Early Palaeozoic age (HANKEL, 1994). The new type section proposed is the interval 3250 - 3636 m in well Tulli-1 (coordinates 44° 21' 45"E 06° 18' 10"N). The lacustrine sediments of this interval are underlain by the Calub Sandstone (3636 - 3958 m) and overlain by the Gumburo Sandstone (2808 - 3250 m).

An unpublished palynological study indicates a Late Permian age for the upper part of the Calub Sandstone, an Early Triassic age for the Tulli Shale, a Late Triassic age for the Gumburo Sandstone and an Early Jurassic age for the Adigrat Sandstone. The microfloral assemblages recorded in this previous study have been correlated with the *Playfordiaspora crenulata* Zone, the *Lunatisporites pellucidus* Zone, the lowert part of the *Triplexisporites playfordii* Zone and the lower and upper part of the *Corollina torosa* Zone of Australia (HANKEL, 1995a, b, c). The present paper deals with palynomorph assemblages from the Adigrat Sandstone.

Source of material

The material studied comes from well Abred-1 (coordinates 45° 14' 53"E 05° 30' 02"N) which was drilled by Gewerkschaft Elwerath in 1963 in the southeastern Ogaden region of Ethiopia. The core material is stored in the magazine of ExxonMobil Production Deutschland GmbH in Nienhagen near Celle.

The well reached a total depth of 3104.40 m. The drill floor elevation was 523 m above sea level. Arenaceous sediments drilled from surface to a depth of 170 m have been assigned to the Jesomma Sandstone. The bore then penetrated the marine Jurassic-Cretaceous succession (170 - 2950 m) and the underlying Adigrat Sandstone (2950 - 3070 m). Well Abred-1 bottomed finally in high grade metamorphic basement rocks (3070 - 3104.40 m) (ELWERATH, 1964).

For palynological analysis seven shale samples have been collected from a cored interval (3007.10 - 3023.80 m) in the middle part of the Adigrat Sandstone. The palynomorph yield of three samples from the lower part of the core was low. The other four samples from the upper part of the core, however, yielded rich and well-preserved palynomorph assemblages on which this study is based. These samples come from a depth of 3010.10 m, 3010.20 m, 3010.25 m, and 3014.90 m.

Techniques and storage of material

The samples have been processed using a standardized palyno-logical technique as described in HANKEL (1991, p. 130). A number of strew slides were prepared in glycerine for each sample. After identification and counting of palynomorphs representative specimens have been picked from these slides for the preparation of single grain mounts (see Appendix). All specimens were photographed and are documented on film negatives. The material (remainders of the shale samples, maceration residues, strew slides, single grain mounts, and film negatives) is held at the author´s collection (catalogue nos. ET-191190 and ET-90-AB-1).

List of taxa identified

A   SPORES
A 1   TRILETE SPORES
A 1.1 AZONATE FORMS
Anapiculatisporites Potonié & Kremp 1954
   Anapiculatisporites dawsonensis Reiser & Williams 1969
Calamospora Schopf, Wilson & Bentall 1944
   Calamospora tener (Leschik) de Jersey 1962
Ceratosporites Cookson & Dettmann 1958
   Ceratosporites helidonensis de Jersey 1971
Deltoidospora Miner emend. Potonié 1956
   Deltoidospora directa (Balme & Hennelly) Norris 1965
Dictyophyllidites Couper emend. Dettmann 1963
   Dictyophyllidites harrisii Couper 1958
   Dictyophyllidites mortonii (de Jersey) Playford & Dettmann 1965

Granulatisporites Ibrahim emend. Potonié & Kremp 1954
  Granulatisporites sp.
Leptolepidites Couper emend. Norris 1968
  Leptolepidites sp.
Lophotriletes (Naumova) Potonié & Kremp 1954
  Lophotriletes sp.
Neoraistrickia Potonié 1956
  Neoraistrickia sp. A of Filatoff 1975
Osmundacidites Couper 1953
  Osmundacidites wellmanii Couper 1953
Retitriletes Pierce emend. Döring, Krutzsch, Mai & Schulz 1963
  Retitriletes semimuris (Danzé-Corsin & Laveine) McKellar 1974
  Retitriletes spp.
Todisporites Couper 1958
  Todisporites sp.
Trachysporites Nilsson 1958
  Trachysporites fuscus Nilsson 1958
A 1.2 CINGULATE FORMS
Antulsporites Archangelsky & Gamerro 1966
  Antulsporites sp.
Foraminisporis Krutzsch 1959
  Foraminisporis tribulosus Playford & Dettmann 1965
Foveosporites Balme 1957
  Foveosporites sp.
Limatulasporites Helby & Foster in Foster 1979
  Limatulasporites limatulus (Playford) Helby & Foster in Foster 1979
Nevesisporites de Jersey & Paten 1964
  Nevesisporites vallatus de Jersey & Paten 1964
Polycingulatisporites Simoncsics & Kedves 1961
  Polycingulatisporites mooniensis de Jersey & Paten 1964
Stereisporites Pflug in Thomson & Pflug 1953
  Stereisporites spp.
A 1.3 CAVATE FORMS
Densoisporites Weyland & Krieger emend. Dettmann 1963
  Densoisporites psilatus (de Jersey) Raine & de Jersey in Raine, de Jersey
  & Ryan 1988
A 2 ALETE SPORES
Pilasporites Balme & Hennelly 1956
  Pilasporites marcidus Balme 1957

B   POLLEN
B 1   MONOSACCATE FORMS
Callialasporites Sukh Dev 1961
   Callialasporites turbatus (Balme) Schulz 1967
B 2   DISACCATE FORMS
Alisporites Daugherty 1941
   Alisporites australis de Jersey 1962
Platysaccus Naumova ex Ishchenko emend. Potonié & Klaus 1954
   Platysaccus queenslandi de Jersey 1962
Podocarpidites Cookson emend. Potonié 1958
   Podocarpidites ellipticus Cookson 1947
B 3   INAPERTURATE FORMS
Araucariacites Cookson ex Couper 1953
   Araucariacites australis Cookson 1947
Inaperturopollenites Thomson & Pflug emend. Potonié 1958
   Inaperturopollenites nebulosus Balme 1970
B 4   MONOSULCATE FORMS
Chasmatosporites Nilsson emend. Pocock & Jansonius 1969
   Chasmatosporites sp.
Cycadopites Wodehouse ex Wilson & Webster 1946
   Cycadopites follicularis Wilson & Webster 1946
Retisulcites Scheuring 1970
   Retisulcites sp.
B 5   RIMULATE FORMS
Classopollis Pflug 1953
   Classopollis chateaunovi Reyre 1970
   Classopollis meyeriana (Klaus) de Jersey 1973
   Classopollis simplex (Danzé-Corsin & Laveine) Reiser & Williams 1969
B 6   MONOPORATE FORMS
Perinopollenites Couper 1958
   Perinopollenites elatoides Couper 1958

C   ACRITARCHS
C 1   SPHAEROMORPHS
Leiosphaeridia Eisenack 1958
   Leiosphaeridia sp.

Composition

The palynomorph assemblages recovered are of corresponding composition (Table I) and consist essentially of pteridophytic and bryophytic spores and gymnosperm pollen associated with small numbers of acritarchs.

Spores make up only a minor proportion of the miospore suite (4 - 7%, average 6%) and are represented mainly by acavate trilete forms. Most abundant are specimens which in this paper have been assigned to the *Neoraistrickia* sp. A / *Retitriletes semimuris* complex (cf. FILATOFF, 1975, p. 53).

The miospore suite is dominated by gymnosperm pollen (93 - 96%, average 94%). Predominant element is the genus *Classopollis* (92 - 95%, average 93%). This taxon is known to be produced by thermophilic and xerophytic conifers of the family Cheirolepidiaceae.

Sphaeromorph acritarchs assignable to *Leiosphaeridia* sp. constitute between 2 - 4% of the total assemblage (average 3%).

Correlation

The prominence of *Classopollis* and the presence of taxa as *Anapiculatisporites dawsonensis, Ceratosporites helidonensis, Dictyophyllidites harrisii, D. mortonii, Foraminisporis tribulosus, Nevesisporites vallatus, Retitriletes semimuris, Alisporites australis, Araucariacites australis, Callialasporites turbatus,* and *Perinopollenites elatoides* demonstrate a close relationship with palynofloras of the *Corollina torosa* Zone of Australia. This zone is of Hettangian to earliest Toarcian age (HELBY et al., 1987; PARTRIDGE, 2006).

Table I

Counts of palynomorph taxa in samples studied

Taxa	Sample depth [m]			
	3014.90	3010.25	3010.20	3010.10
SPORES (total)	36	24	28	35
Anapiculatisporites dawsonensis	+	–	2	1
Antulsporites sp.	+	+	1	+
Calamospora tener	+	+	–	–
Ceratosporites helidonensis	1	–	+	4
Deltoidospora directa	1	2	+	2
Densoisporites psilatus	–	–	+	–
Dictyophyllidites harrisii	+	+	–	–
Dictyophyllidites mortonii	1	2	+	+
Foraminisporis tribulosus	–	–	1	1
Foveosporites sp.	–	–	+	–
Granulatisporites sp.	–	–	+	–
Leptolepidites sp.	+	–	–	–
Limatulasporites limatulus	1	+	+	+
Lophotriletes sp.	+	1	–	+
Neoraistrickia sp. A / *Retitriletes semimuris* complex	13	6	14	14
Nevesisporites vallatus	–	+	1	–
Osmundacidites wellmanii	+	+	1	1
Pilasporites marcidus	1	–	+	1
Polycingulatisporites mooniensis	–	–	–	+
Retitriletes spp.	–	+	+	+
Stereisporites spp.	1	+	3	3
Todisporites sp.	1	–	+	–
Trachysporites fuscus	–	–	–	1
unidentified trilete spores	16	13	5	7
POLLEN (total)	486	522	490	508
Alisporites australis	+	+	3	1
Araucariacites australis	3	4	1	+
Callialasporites turbatus	1	+	7	+
Chasmatosporites sp.	–	–	+	–
Classopollis spp.	480	518	477	507
Cycadopites follicularis	2	–	–	+
Inaperturopollenites nebulosus	–	+	1	–
Perinopollenites elatoides	+	–	+	–
Platysaccus queenslandi	–	–	1	–
Podocarpidites ellipticus	+	–	1	+
Retisulcites sp.	–	–	–	+
Total miospore count	522	546	519	543
ACRITARCHS				
Leiosphaeridia sp.	21	13	14	9
Total palynomorph count	543	559	533	552

– = not observed; + = observed but not in actual count

Microfloral assemblages from the Adigrat Sandstone in the Ogaden region have been recorded previously from wells Bokh-1, Tulli-1, Hilala-4, and Calub-3 (HANKEL, 1994, 1995a, b, c). Based on compositional differences a provisional two-fold zonation of these assemblages has been proposed. Palynofloras from the middle part of the Adigrat Sandstone are characterized by high frequencies of *Classopollis* (83 - 99%), the presence of *Perinopollenites elatoides* (up to 5%), and the non-presence of *Exesipollenites tumulus*. These older assemblages have been correlated with the lower part of the *Corollina torosa* Zone of Australia (HANKEL, 1995b). In contrast, palynofloras from the upper part of the Adigrat Sandstone are characterized by lower frequencies of *Classopollis* (60 - 68%), the presence of *Exesipollenites tumulus* (up to 5%), and the non-presence of *Perinopollenites elatoides*. These younger assemblages have been correlated with the upper part of the *Corollina torosa* Zone of Australia (HANKEL, 1995b).

The foregoing explanations suggest a correlation of the present palynoflora with those of the lower part of the *Corollina torosa* Zone of Australia.

Conclusions

(1) The middle part of the Adigrat Sandstone in well Abred-1 is of Hettangian to Sinemurian age.
(2) The sediments of the studied section were deposited in a continental environment under hot and dry climatic conditions.
(3) The presence of small numbers of leiospheres indicates a restricted marine influence.
(4) The further use of the term Bokh Shale for Early Triassic lacustrine deposits in the subsurface of the Ogaden Basin should be abandoned.

Acknowledgements

BEB Erdgas und Erdöl GmbH & Co. KG (Hannover) has granted access to core material of well Abred-1. The author gratefully acknowledges the support.

References

ELWERATH, 1964: Schichtenverzeichnis der Aufschlußbohrung Abred 1. 75 pp., unpublished report.

FILATOFF, J., 1975: Jurassic palynology of the Perth Basin, Western Australia. Palaeontographica B, 154, 1-113.

HANKEL, O., 1991: Early Triassic plant microfossils from the Kavee Quarry section of the Lower Mariakani Formation, Kenya. Rev. Palaeobot. Palynol., 68, 127-145.

HANKEL, O., 1994: Palynology of core samples from the lower part of well Bokh-1, Ogaden Basin, Ethiopia. 14 pp., unpublished report.

HANKEL, O., 1995a: Palynology of core samples from the lower part of well Tulli-1, Ogaden Basin, Ethiopia. 28 pp., unpublished report.

HANKEL, O., 1995b: Palynology of core samples from the lower part of well Hilala-4, Ogaden Basin, Ethiopia. 23 pp., unpublished report.

HANKEL, O., 1995c: Palynology of core samples from the lower part of well Calub-3, Ogaden Basin, Ethiopia. 41 pp., unpublished report.

HELBY, R., MORGAN, R. & PARTRIDGE, A.D., 1987: A palynological zonation of the Australian Mesozoic. Mem. Assoc. Australas. Palaeontol., 4, 1-94.

PARTRIDGE, A.D., 2006: Jurassic – Early Cretaceous spore-pollen and dinocyst zonations for Australia. In: MONTEIL, E. (coordinator), Australian Mesozoic and Cenozoic palynology zonations – updated to the 2004 Geologic Time Scale. Geoscience Australia Record 2006/23.

WORKU, T., 1988: Sedimentology, diagenesis and hydrocarbon potential of the Karoo sediments (Late Palaeozoic to Early Jurassic), Ogaden Basin, Ethiopia. Thesis, Postgraduate Research Institute for Sedimentology, University of Reading, 222 pp.

Appendix

Register of palynomorph species preserved as single grain mounts

Species	Single grain mount	Overall size [μm]	Film negative
Alisporites australis	3b-11	55	I-0-3
	17-10	65	I-4-5
Anapiculatisporites dawsonensis	3b-7	43	I-6-7
	17-29	42	I-8-9
Antulsporites sp.	3a-9	36	I-10-11
	3a-33	40	I-12-13
	3b-12	41	I-14-15
	8-13	46	I-16-17
Araucariacites australis	3b-1	83	I-18-19
	17-38	60	I-20-21
Calamospora tener	3b-16	33	I-22-23
	8-23	37	I-24-25
Callialasporites turbatus	3a-1	52	I-26-27
	17-39	58	I-28-29
Ceratosporites helidonensis	3a-23	28	I-30-31
	17-41	28	I-32-33
Chasmatosporites sp.	3a-35	42	I-34-35
Classopollis chateaunovi	3a-5	32	I-36-37
	17-25	35	II-1-2
	17-26	36	II-3-4
	17-31	28	II-5-6
	17-34	30	II-7-8
Classopollis meyeriana	8-8	30	II-9-10
	17-13	31	II-11-12-13
Classopollis simplex	3b-30	28	II-14-15
	17-48	23	II-16-17
Cycadopites follicularis	8-28	37	II-18-19
	17-44	43	II-20-21
Deltoidospora directa	3a-2	33	II-22-23
	8-16	27	II-24-25
	8-30	45	II-26-27
Densoisporites psilatus	3a-6	40	II-28-29
Dictyophyllidites harrisii	3b-24	47	II-30-31
	8-27	49	II-32-33
Dictyophyllidites mortonii	3a-44	35	II-34-35-36
	17-46	33	III-1-2
Foraminisporis tribulosus	3a-42	37	III-3-4
	17-8	38	III-5-6
Foveosporites sp.	3a-28	30	III-7-8-9
Granulatisporites sp.	3a-26	37	III-10-11

Appendix (continued)

Inaperturopollenites nebulosus	3a-4	140	III-12-13-15
	3b-15	82	III-16-17-18-19
Leiosphaeridia sp.	3b-3	33	III-20-21
	3b-10	33	III-22-23
Leptolepidites sp.	8-15	35	III-24-25
	8-22	32	III-26-27
Limatulasporites limatulus	3a-24	43	III-28-29-30
	17-16	40	III-31-32
Lophotriletes sp.	8-18	68	III-33-34
	17-49	54	III-35-36
Neoraistrickia sp. A	3a-25	38	IV-1-2
	3a-29	43	IV-3-4
	17-12	37	IV-5-6
Nevesisporites vallatus	3a-38	37	IV-7-8
	3b-28	45	IV-9-10
Osmundacidites wellmanii	3b-6	45	IV-11-12
	3b-9	45	IV-13-14
Perinopollenites elatoides	3a-43	38	IV-15-16
	8-34	37	IV-17-18
Pilasporites marcidus	3a-31	41	IV-19-20
	3a-39	41	IV-21-22
Platysaccus queenslandi	3a-45	68	IV-23-24
Podocarpidites ellipticus	8-29	48	IV-25-26
	8-32	63	IV-27-28
	17-35	50	IV-29-30
Polycingulatisporites mooniensis	17-50	33	IV-31-32
Retisulcites sp.	17-47	36	IV-33-34
Retitriletes semimuris	3a-12	41	IV-35-37-38
	3a-18	38	V-1-2
	3a-27	42	V-3-4
Retitriletes spp.	3a-41	58	V-5-6
	3b-21	56	V-7-8
Stereisporites spp.	3a-7	38	V-9-10
	3a-22	30	V-11-12
	3b-13	28	V-13-14
	8-4	35	V-15-16
	8-24	40	V-17-18
	17-17	35	V-19-20
Todisporites sp.	3a-3	45	V-21-22
	8-26	45	V-23-24
Trachysporites fuscus	17-45	41	V-25-26

The number of single grain mounts is divided into two parts. The first part denotes the sample number (17 = depth 3010.10 m; 3a = depth 3010.20 m; 3b = depth 3010.25 m; 8 = depth 3014.90 m).